CARING FOR MILK SNAKE

UNDERSTANDING AND TAKING CARE OF YOUR MILK SNAKE

Practical Advice for Owners and Behavior

DR MORRIS HART

Copyright© 2024 **DR MORRIS HART**

All rights reserved. No part or part of this book or publication may be reproduced, stored, or transferred in any form by electronic, mechanical, recording, or other retrieval system without written permission from the publisher

Table of Contents

INTRODUCTION 5

CHAPTER 1 11

SPECIES & VARIETIES TO CONSIDER WHEN CHOOSING THE IDEAL MILK SNAKE 11

CHAPTER 2 19

CREATING THE PERFECT HABITAT: ENVIRONMENTAL REQUIREMENTS AND ENCLOSURES 19

CHAPTER 3 27

KNOWING MILK SNAKE BEHAVIOR: FROM EATING TO GETTING ALONG WITH OTHERS 27

CHAPTER 4 36

TAKING CARE OF YOUR PET AND DEVELOPING A TRUSTING RELATIONSHIP 36

CHAPTER 5 45

Health and Well-Being: Keeping Your Milk Snake Healthy 45

CHAPTER 6 59

Typical Problems and Their Fixes: Resolving Pet Care Concerns 59

CHAPTER 7 69

Fun Exercises and Enhancement: Maintaining the Stimulation of Your Milk Snake 69

CHAPTER 8 82

Common Questions and Answers (FAQs) about Milk Snakes 82

CHAPTER 9 91

In conclusion, relishing the special pleasures of owning a milk snake 91

Introduction

Milk snakes are fascinating reptiles that are under the genus Lampropeltis. They are renowned for their vivid colors, kind disposition, and low maintenance needs. These non-venomous constrictors, which are native to North, Central, and South America, have won the hearts of reptile lovers everywhere. We will dig into the intriguing world of milk snakes in this thorough book, learning about their natural history, traits, and the pleasures and responsibilities of owning them as pets.

- The Distribution and Natural History of

Milk snakes are members of the Colubridae family, which includes a wide variety of snake species. There are about 25 species in the genus Lampropeltis that are known to exist, the most well-known of which being the milk snake (Lampropeltis triangulum). Depending on the species and subspecies, these snakes can live in a range of environments, such as meadows, woods, rocky hillsides, and even cities.

Milk snakes are native to the Americas and can be found throughout much of the continent, from southern South America to southeast Canada. They are found, among other places, in the US, Mexico, Guatemala, Honduras, Costa Rica, and Venezuela. They can survive in a variety of habitats, including tropical rainforests and temperate woods, thanks to their adaptability.

- Physical attributes

Across species and subspecies, milk snakes exhibit a wide range of vivid patterns and colors that contribute to their remarkable look. They often have an elongated, thin body with glossy, smooth scales. Their coloring usually comprises of red, black, and yellow broad bands or blotches, though the precise pattern varies based on heredity and geography.

The "tri-color" pattern, which has alternate bands of red, black, and yellow or white and resembles the deadly coral snake (Micrurus spp.), is one of the most well-known varieties of the milk snake pattern. By mimicking the look of a dangerous species, this mimicry, often referred to as Batesian mimicry, acts as a type of protective adaptation, keeping prospective predators at bay.

- Conduct and Attitude

As placid and tolerant of handling, milk snakes are a popular choice for reptile aficionados of all skill levels, despite their colorful look. If given the right attention and socialization, they can get very acclimated to being around people and behave calmly during handling sessions.

Since milk snakes are generally nocturnal or crepuscular, the evening and early morning hours are when they are most active. They may hide throughout the day in isolated areas such as rock crevices, underground tunnels, or other hidden places to ward off predators and maintain body warmth.

A wide range of tiny creatures, including rodents, birds, reptiles, and amphibians, are the prey of milk snakes, which hunt and feed like opportunistic predators. They happily consume commercial rodent prey, like mice and rats, when kept in captivity, but it's important to feed them meals that are the right size to avoid obesity and other health problems.

- Status of Conservation

Although milk snakes are not regarded as an endangered or threatened species, habitat loss, fragmentation, and human encroachment may pose limited hazards to some populations. A few populations of milk snakes are also declining as a result of human persecution and illicit capture for the pet trade.

Milk snakes and other reptile species depend on conservation initiatives that safeguard their habitat, uphold laws prohibiting the illegal trade, and encourage ethical pet ownership.

- The Pleasure of Owning Pet Milk Snakes

Milk snakes' compelling beauty, manageable size, and relatively minimal maintenance requirements make them appealing pets for many reptile aficionados. In contrast to many larger snake species, milk snakes usually grow to a manageable length of two to four feet, so people with limited space or experience can safely handle them.

Furthermore, milk snakes thrive in well-kept captivity settings with ideal humidity, temperature, and habitat furnishings because they are typically resilient and adaptive animals. They can survive in captivity for well over ten years with the correct care and attention, giving their owners years of company and pleasure.

Milk snakes are amazing reptiles that captivate people all around the world with their distinct attractiveness and rich natural history. Pet owners of all ages and skill levels can enjoy a gratifying experience with these snakes because of their brilliant colors, placid nature, and adaptability in captivity. You can have a rewarding experience owning a milk snake if you recognize their needs and give them the attention and stimulation they require.

Chapter 1

Species & Varieties to Consider When Choosing the Ideal Milk Snake

Selecting the ideal milk snake for your house is a thrilling but crucial choice. With such a wide variety of species and types to choose from, each with distinct traits and maintenance needs, it's important to do your homework and weigh all the options before choosing. We will examine the various types and varieties of milk snakes in this extensive guide, offering insights to assist you in selecting the ideal companion for your tastes and way of life.

- Knowing the Species of Milk Snakes

Milk snakes are members of the Lampropeltis genus, which is home to many subspecies and several recognized species. The majority of milk snakes have similar basic care needs, however they vary greatly in size, color, and preferred natural environment. Let's examine some of the most well-known species of milk snakes in more detail:

The eastern subspecies of the milk snake (Lampropeltis triangulum triangulum), which is native to Canada and the eastern United States, is distinguished by its unique red, black, and yellow banding pattern. Because of their adaptability and relative docility, eastern milk snakes are great options for novice reptile enthusiasts.

Mexican Milk Snake (Lampropeltis triangulum annulata): With vivid red, black, and white patterning, Mexican milk snakes are native to Mexico and some regions of Central America. Compared to their eastern counterparts, they are typically a little bigger and may have greater variation in color.

The Pueblan Milk Snake, or Lampropeltis triangulum campbelli, is a native of central Mexico's highlands. It is distinguished by its eye-catching bands of red, black, and white. Since they are smaller than some other species of milk snakes, they are well-liked by reptile lovers who have limited room.

The Sinaloan Milk Snake (Lampropeltis triangulum sinaloae) is a type of milk snake that is native to western Mexico. Its striking stripes are red, black, and white. They are well-suited as pets for both novice and seasoned keepers because to their versatility and peaceful disposition.

Nelson's Milk Snake (Lampropeltis triangulum nelsoni): Native to Mexico's mountainous highlands, Nelson's milk snakes have striking banding patterns in red, black, and white. Because of their natural habitat, they may benefit from slightly colder temperatures even though they require care similar to other species of milk snakes.

- Things to Take Into Account While Choosing a Milk Snake

A number of considerations should be made while selecting a milk snake for your house in order to guarantee the health and compatibility of you and your new companion. Think about the following elements:

Size: Although the majority of milk snakes are small in comparison to other snake species, there can be a considerable difference in size between various subspecies. Establish the mature size of the species you are interested in and make sure you have enough room and amenities for it.

Although milk snakes are usually calm and amenable to handling, temperaments can differ throughout individuals. Before deciding how to evaluate the snake's mood and behavior, try to engage with it.

Color and Design: Milk snakes are well-known for their vivid hues and striking designs, which can differ greatly between species and subspecies. Take your aesthetic tastes into account and select a snake whose colors and patterns please you.

Availability: Certain species of milk snakes could be easier to find as pets than others. To locate a healthy and properly obtained snake, check for trustworthy online retailers, local breeders, and reptile expos.

Care Requirements: In terms of temperature, humidity, and habitat furnishings, different species of milk snakes may have slightly varied needs. Make sure you are equipped to handle the unique requirements of the species you select.

- Popular Types of Milk Snakes

There are other color and pattern variations of milk snakes that have been selectively developed in captivity, in addition to the main species and subspecies. These mutations add to the variety of alternatives accessible to potential owners by frequently displaying distinctive and eye-catching traits. Popular variants of milk snakes include:

Albino: Albino milk snakes have a white or cream background with pink or red markings due to a lack of melanin pigment. Because of its remarkable appearance, this morph is extremely sought for.

Hypo: A genetic mutation known as hypomelanism causes the skin of a snake to become lighter in color and more visually appealing by reducing the amount of black pigment. Hypo milk snakes frequently have vivid orange and red colors.

Anerythristic: Milk snakes that are anerythristic lack red pigment, giving them a coloring of black, white, and gray. The look of this morph can vary; some individuals may have highly contrasted banding, while others may have more muted hues.

Amelanistic: Amelanistic milk snakes have red and yellow pigmentation but no black pigment, much as albino variants. All in all, they seem paler than milk snakes of the wild variety.

Scaleless: The absence of scales on the body of scaleless milk snakes gives them a distinct, smooth appearance. Due to its rarity, this morph might need specific attention because it lacks protective scales.

A rewarding and fulfilling experience for you and your new pet is guaranteed when you carefully evaluate species, types, and individual traits when choosing the ideal milk snake. There is a milk snake out there to fit every taste and lifestyle, whether you're drawn to the vivid colors of wild-type milk snakes or the distinctive patterns of morphs created in captivity. By doing your homework and learning about your options, you'll be ready to make an informed choice and dive into the exciting world of owning a milk snake.

Chapter 2

Creating the Perfect Habitat: Environmental Requirements and Enclosures

For the sake of your milk snake's general health and happiness, you must provide the ideal habitat. Every element of your snake's living habitat, from selecting the ideal enclosure to creating the ideal atmosphere, is vital to its well-being. We will go over the essential elements of creating the perfect home for your milk snake in this in-depth tutorial, so that it can flourish in its new surroundings.

- Choosing an Enclosure

The choice of a suitable enclosure is the first stage in creating a habitat for your milk snake. When selecting a habitat for your snake, take into account the following factors:

Size: Milk snakes need lots of room to roam about and investigate their surroundings. Generally speaking, the enclosure need to be at least twice as long as the snake, giving it ample space for hiding and climbing.

Material: Popular options for housing milk snakes include plastic enclosures and glass terrariums. While plastic enclosures offer insulation and humidity retention, glass offers superior visibility.

Ventilation: To preserve air quality and avoid the accumulation of moisture and smells, make sure the enclosure has enough ventilation. Ventilation panels or screened tops can aid in improving airflow.

Security: To keep your snake safe and prevent escapes, choose an enclosure with strong locking systems. Since snakes are skilled escape artists, it's imperative to get a trustworthy locking system.

Accessibility: Choose an enclosure that is simple to clean and maintain. For ease of use and security, sliding doors or front-opening enclosures are preferred.

- Furnishings for Habitation

After selecting the cage, it's time to fill it with all the things your milk snake needs to live in a cozy and stimulating environment. Here are some crucial furnishings for your habitat to think about:

Select a substrate that will give your milk snake a comfortable surface for burrowing and exploring while also emulating its natural habitat. Aspen shavings, cypress mulch, coconut husk, or paper-based bedding are examples of appropriate substrate choices.

Hiding Places: Due to their natural tendency toward secrecy, milk snakes need hiding places to feel safe. To suit your snake's predilection for different temperatures, provide it with at least two hiding places, one on the warm side and one on the cool side of the cage.

Climbing Equipment: Because milk snakes are semi-arboreal, they can value having climbing equipment in their enclosure. Your snake can find enrichment and climbing surfaces in branches, logs, and artificial vines.

Water Bowl: Give your milk snake access to a shallow water bowl so it can drink and relax. To avoid tipping, use a heavy, stable bowl and make sure it gets cleaned and filled with fresh water on a regular basis.

Heating and Lighting: Milk snakes are ectothermic reptiles, which means that they cannot control their body temperature without the assistance of outside heat sources. To produce a thermal gradient inside the enclosure, provide a heat source, such as a heat lamp or under-tank heating pad. You should also think about using a full-spectrum UVB light to provide your snake the UV sunlight it needs to stay healthy.

- Conditions for Temperature and Humidity

It's essential to keep your milk snake at the right temperature and humidity levels for its overall health and wellbeing. To guarantee ideal environmental conditions, adhere to these guidelines:

Temperature Gradient: By placing a heat source on one side and a colder area on the other, you can create a thermal gradient inside the enclosure. The ideal temperature ranges for the heated side are 85–90°F (29–32°C) while the cool side is 75–80°F (24–27°C). Regularly check temperature levels with a thermometer.

Temperature Drop at Night: To replicate natural conditions, allow for a small temperature drop at night.

Throughout the enclosure, nighttime temperatures can safely fall between 70 and 75°F (21 and 24°C).

Humidity Levels: To maintain proper shedding and respiratory function, milk snakes need a modest amount of humidity. The habitat should have a 40–60% humidity level, which can be attained by misting, offering a humid hide, or employing a humidifier for reptiles.

Monitoring and Modifying: To keep an eye on humidity levels and make any modifications, use a hygrometer. If the humidity is excessively high, lower it with a dehumidifier or enhance ventilation. To raise humidity, spritz the enclosure or add moisture-retaining substrate.

- Enhancement of Environment

Milk snakes benefit from habitat enrichment to stimulate their natural behaviors and enhance general well-being, in addition to their fundamental needs for shelter, warmth, and humidity. Take into account these suggestions for habitat enrichment for your snake:

Climbing structures: To help your milk snake explore and use its innate climbing skills, provide branches, logs, and other climbing structures.

Burrowing Substrate: Provide a thick layer of substrate so your snake may tunnel and burrow, much like it would in the wild.

Feeding Enrichment: Use techniques for feeding enrichment, include placing items of prey in various areas or simulating hunting behavior with feeding tongs.

Sensory Stimulation: To arouse your snake's senses and offer mental stimulation, add new smells, textures, and items to the enclosure.

Even though milk snakes are solitary creatures, on occasion handling and engagement with their owner can foster social enrichment and the development of confidence and trust.

For the sake of your milk snake's general well-being and general enjoyment, you must set up the perfect home. You may create a cozy and stimulating environment for your snake to live in by choosing the right enclosure, filling it with necessities for the habitat, and keeping the temperature and humidity at ideal levels. Furthermore, adding environmental enrichment techniques to your snake's habitat will encourage its innate habits and enhance both its physical and mental health. You can make sure your snake has a long and happy life in captivity by giving it the care and attention it requires, including taking note of its environmental requirements.

Chapter 3

Knowing Milk Snake Behavior: From Eating to Getting Along with Others

Like all reptiles, milk snakes display a variety of behaviors that are impacted by their innate instincts, their surroundings, and their particular temperaments. Comprehending the behavior of milk snakes is crucial for ensuring their best care and enrichment in captivity, encompassing everything from eating and digesting to thermoregulation and sociability. We will examine several facets of milk snake life in this thorough guide, including feeding patterns, thermoregulatory behavior, mating practices, and social interactions.

- Feeding Patterns

Carnivorous reptiles, milk snakes consume a wide range of tiny creatures, including rodents, birds, amphibians, and reptiles. They hunt by sitting and waiting in the woods, using ambush and stealth to bring down their prey. On the other hand, when provided pre-killed or frozen and thawed prey by their masters, they gladly take it.

Appropriately sized prey items are crucial while feeding milk snakes in order to avoid regurgitation and other digestive problems. While adult snakes can eat larger prey items like adult mice or small rats, hatchlings and juvenile snakes should be fed smaller prey items like fuzzy or pinky mice.

The age, size, and metabolism of the snake all affect how frequently it needs to be fed; younger snakes normally need to eat more frequently than adult snakes. Adults can be fed every 7–10 days, however hatchlings and juveniles should generally be fed every 5–7 days.

- Metabolism and Digestion

Following a meal, milk snakes go through a digestive process that breaks down their prey and allows their bodies to absorb nutrients. Strong stomach acids and enzymes aid in digestion, enabling snakes to effectively absorb nutrients from their diet.

Snakes may get drowsy during digesting and look for warm areas in their habitat to help with the process. By creating a thermal gradient inside the enclosure, snakes are able to control their body temperature and improve their digestive processes.

Snake metabolism is impacted by a number of variables, including temperature, activity level, and frequency of feeding. Compared to mammals, snakes have a slower metabolism, which enables them to go without food for long periods of time. However, to maintain the best possible health and metabolic function, regular feeding and good husbandry are necessary.

- Behavior Thermoregulatory

Thermoregulation, the method by which milk snakes control their body temperature, is a requirement shared by all reptiles. Snakes are able to maintain their ideal body temperature and metabolic rate by alternating between warm and cool regions in their habitat.

A thermal gradient with a warm side and a cool side must be provided in a confinement enclosure. Snakes can now thermoregulate by relocating within the enclosure as needed. It is recommended to keep the cold side at 75-80°F (24-27°C) and the warm side at approximately 85-90°F (29-32°C).

In order to increase their body temperature, snakes frequently bask beneath heat lamps or heat pads, particularly after eating. Snakes can effectively regulate their body temperature and preserve their general health and well-being when given a diverse range of temperature gradients and hiding places.

- Breeding Patterns

As oviparous reptiles, milk snakes procreate by laying eggs. When temperatures are higher and daylight hours are longer, milk snakes tend to breed in the spring or early summer. Male snakes may become more active and aggressive during this period as they compete for mates.

Although milk snake courting behavior differs by species, ritualized displays like male fights or courtship dances are frequently used. After mating, females will go through a few weeks of gestation before releasing a clutch of eggs.

Milk snake rearing in captivity necessitates close monitoring of environmental factors such as temperature, humidity, and photoperiod. Breeding behavior and egg production can be encouraged by simulating a winter cold phase and then gradually increasing the temperature and humidity.

- Getting Along and Managing

With the right socialization, milk snakes, like many other reptiles, can grow accustomed to being handled and interacting with their owners. Snakes can become less stressed during handling sessions and become desensitized to human contact with regular, gentle handling starting at an early age.

It's crucial to support a milk snake's body correctly when working with it; abrupt movements or rough treatment should be avoided. If handled incorrectly, snakes might get protective or anxious, so it's important to approach them quietly and confidently.

Although milk snakes are generally calm and amenable to handling, each one's personality is unique. Certain snakes may exhibit higher levels of anxiety or defensiveness than others, particularly if they are not used to being handled frequently. It's critical to treat your snake with caution and patience, and to respect its boundaries.

- Enhancement of Behavior

Milk snakes gain from environmental enrichment in addition to receiving their fundamental needs of food, water, and shelter since it encourages their natural behaviors and enhances their mental and physical health. Among the enrichment activities are:

Give your snake the chance to explore its enclosure and find new hiding places, climbing frames, and sensory elements.

Feeding Enrichment: Use techniques for feeding enrichment, include presenting prey items in various areas of the enclosure or simulating hunting behavior with feeding tongs.

Sensory Stimulation: To arouse your snake's senses and offer mental stimulation, add new smells, textures, and items to the enclosure.

Even though milk snakes are solitary creatures, on occasion handling and engagement with their owner can foster social enrichment and the development of confidence and trust.

To give milk snakes in captivity the best care and enrichment possible, one must comprehend their behavior. Every facet of behavior, from eating and digestion to thermoregulation and sociability, is vital to the health and welfare of these amazing reptiles. You can make sure your snake has a happy and healthy life in captivity by paying attention to its behavior and adapting accordingly.

Chapter 4

Taking Care of Your Pet and Developing a Trusting Relationship

It's a fulfilling experience to handle and form a bond with your milk snake; it can help your pet socialize and become healthier while also strengthening your relationship with it. Building trust and confidence with your snake is crucial for its care and your happiness as a pet owner, regardless of your level of experience with reptiles. We will go over the fundamentals of caring for and developing a deep association with milk snakes in this extensive book, along with some helpful hints and methods to get you started.

- Comprehending the Behavior of Snakes

Prior to learning handling skills, it's important to comprehend milk snake behavior and environment perception. Snakes are ectothermic reptiles with a distinct range of activities and sensory capacities, including milk snakes:

Sensory Perception: Although they have weak eyesight, snakes have well developed senses of vibration and smell. They use tongue flicking to collect scent particles from the atmosphere to identify possible mates, predators, and prey.

Milk snakes may display protective responses including hissing, flattening their bodies, or striking when they feel threatened or under stress. You can better read your snake's emotions and react accordingly during handling sessions if you are aware of these characteristics.

Exploratory Behavior: Milk snakes are naturally inquisitive and may slither, climb, and investigate novel objects to learn more about their surroundings.

Providing them with chances to explore can aid in mental stimulation and assist keep them from being bored.

- Getting Ready for Sessions on Handling

To reduce stress for both you and your pet, it's crucial to establish a peaceful and controlled environment before attempting to handle your milk snake. To get ready for handling sessions, take the following actions:

Select the Appropriate Time: Choose an hour, preferably in the evening or early morning, when your snake is most likely to be awake and attentive. Snakes can be more defensive or sensitive right after eating or during shedding periods, so avoid handling them during these times.

Wash Your Hands: Make sure to properly wash your hands with unscented soap and water to get rid of any food particles, lotion residue, or other materials that your snake might mistake for threats.

Reduce Distractions: Select a peaceful, quiet space away from loud noises, abrupt movements, and other disturbances that could frighten or upset your snake.

Have proper Handling Equipment: During handling sessions, gently support and lead your snake using proper handling equipment, such as a snake hook or snake tongs. Retaining or grabbing your snake with your hands can be stressful and sometimes harmful.

- Managing Methods

To reduce tension and foster confidence, it's critical to handle your milk snake with calmness and assurance. To ensure safe and efficient handling, abide by these guidelines:

Approach Gradually: Before attempting to handle your snake, approach it with assurance and gently, giving it time to see and smell you. Steer clear of abrupt movements and loud noises that could frighten your snake.

Support the Body: To minimize harm and increase comfort when handling your snake, provide appropriate support for its body. Just behind the head, use one hand to support the front third of the snake's body; use the other to support the middle and tail.

Avoid Squeezing: When handling your snake, refrain from squeezing or restricting its body since this may lead to stress and discomfort. Rather, let your snake roam about freely, investigating your hands and arms at its own leisure.

Be Patient: If your snake exhibits any signs of stress or discomfort, treat it with gentleness and patience. As trust and confidence grow, give it some time to get used to handling before progressively lengthening and increasing the frequency of handling sessions.

Look Out for Stress Signs: During handling sessions, observe your snake's behavior and body language. Hissing, shallow breathing, defensive postures, and flight reflexes are some indicators of stress. Return your snake to its enclosure with care if it exhibits signs of stress, and try again at a later time.

- Establishing Bonds and Trust

With your milk snake, developing trust and a strong bond takes time, patience, consistency, and positive reinforcement. To build a solid and happy relationship with your pet, heed these advice:

Handle Often: Handle your milk snake frequently to help it become used to human interaction and gradually develop trust. When your snake gets more comfortable, gradually increase the length and frequency of your handling sessions from brief, gentle beginnings.

Use Positive Reinforcement: Reward your snake for being calm and collected during handling sessions by giving it a favorite food item or giving it polite praise. As a result, handling will become more positively associated with experiences, fostering confidence and trust.

Respect Boundaries: If your snake exhibits signs of stress or discomfort, respect its boundaries and refrain from pressuring it to interact with you. Give your snake the freedom to set the pace of encounters and give them the chance to retreat if necessary.

Offer Enrichment: To pique your snake's curiosity and promote its innate activities, add environmental enrichment to its cage. Climbing structures, hiding places, and sensory stimuli like unfamiliar textures or smells can all fall under this category.

Be Consistent and Patient: It takes time and patience to develop a bond and trust with your milk snake. Take a methodical approach to handling and refrain from hurrying the procedure. Your snake will develop a sense of trust and a strong attachment with you as its caregiver with time and good reinforcement.

Aside from strengthening your relationship with your pet, handling and connecting with a milk snake can also help with socialization and general well-being. You may create a solid and enduring relationship with your snake by learning about its behavior, being ready for handling sessions, employing safe handling methods, and encouraging trust and bonding through positive reinforcement. You can spend many years with your milk snake as a companion if you are consistent, patient, and handle them gently.

Chapter 5

Health and Well-Being: Keeping Your Milk Snake Healthy

Your milk snake's lifespan and quality of life depend on you taking care of its health and well-being. Proactive care is essential to preserving your snake's general health. This includes giving it the right diet and water, keeping an eye out for any symptoms of sickness, and seeking veterinary attention when necessary. We will examine many facets of milk snake health and wellness in this extensive book, providing advice and ideas to help you give your pet the best care possible.

- dietary and feeding

The health and vitality of your milk snake are mostly dependent on a proper diet. Because they are carnivorous reptiles, milk snakes must eat mostly complete prey items like mice and rats, with the rare small bird or reptile. To make sure your snake is getting enough to eat, adhere to these guidelines:

Prey Size: Provide prey items that are suitable for the size, age, and metabolic rate of your snake. While adult snakes can eat larger prey items like adult mice or small rats, hatchlings and juveniles should be fed smaller prey items like fuzzy mice or pinky mice.

Feeding Frequency: Because of their faster growth and higher metabolic rate, younger snakes usually need to be fed more frequently than adults. Adults can be fed every 7–10 days, while hatchlings and juveniles can be fed every 5–7 days.

Variety: To guarantee your snake has a healthy diet and all the nutrients it needs, provide it with a range of prey items. To offer food variety, think about switching between mice, rats, and other small animals on a rotating basis.

Preparing Prey: Make sure your snake is fully thawed and warmed to room temperature before giving it any pre-killed or frozen-thawed prey items. To avoid scent transmission, deliver the prey item to your snake using feeding tongs rather than handling it with your hands.

Monitoring: To make sure your snake is keeping a healthy weight and growth rate, periodically check its bodily condition and weight. To avoid underfeeding or obesity, alter the frequency of feedings and the size of the prey as necessary.

Water and Hydration

Maintaining the health and physiological processes of your milk snake depends on proper hydration. Make sure your snake always has access to a shallow water bowl full of fresh, clean water so it can drink and soak when needed. For optimal hydration, adhere to following guidelines:

Water Quality: To avoid contamination and bacterial growth, change the water in your snake's bowl on a frequent basis. Before adding more fresh water, properly rinse the water bowl after cleaning it with a little soap and water.

Opportunities for Soaking: To help with shedding or thermoregulation, milk snakes can spend time soaking in their water bowl. Make sure the water dish is big enough so the snake can comfortably immerse itself, and you can add more soaking spots as needed.

spraying: To raise humidity levels and give your snake more hydration, consider spraying the enclosure with water if it shows symptoms of dehydration or is having trouble shedding. Spot the cage sparingly using a reptile-safe spray bottle, paying particular attention to the areas where your snake hangs out.

Observing: Keep an eye out for any indications of dehydration in your snake, such as sunken eyes, wrinkles, or decreased activity. See a veterinarian that specializes in reptiles for advice on rehydrating your snake if you think it may be dehydrated.

- Skin Health and Shedding

As they mature, milk snakes, like all reptiles, regularly lose skin. Snakes shed their skin naturally, which is an important process that helps them to get rid of old, dead skin cells and show new, vivid skin underneath. To encourage your milk snake's skin to stay healthy and shed in a healthy manner, try these tips:

Sufficient Humidity: To encourage shedding, keep the humidity in your snake's enclosure at the right amounts. For milk snakes, a humidity level of 40–60% is excellent; during times of shedding, slightly higher levels (60–70%) are recommended to soften the old skin.

Opportunities for Soaking: During times of shedding, give your snake the chance to soak in a bowl of shallow water or in a damp hide. The old skin will become softer and easier to remove thanks to the increasing humidity.

Gentle Handling: To prevent stress and interfere with the shedding cycle, do not handle your snake when it is shedding. Let your snake shed in peace, and don't handle it again until it's finished.

Helpful Hints: If your snake has trouble shedding, you might want to give it a moist hide or spray the enclosure with a little more water to raise the humidity levels.

Keep in mind that forcing the removal of a stuck shed might harm and stress your snake.

Monitoring: Keep a close eye on your snake's shedding process and be alert for any indications of retained shed or trouble shedding. For advice on appropriate care and support, speak with a veterinarian that specializes in reptiles if you notice any problems.

- surroundings

It's critical to maintain suitable environmental conditions in your milk snake's enclosure for its general health and well-being. To guarantee ideal environmental conditions, adhere to these guidelines:

Temperature gradient: Include a warm and a cool side to the enclosure's thermal gradient. It is recommended to keep the cold side at 75-80°F (24-27°C) and the warm side at approximately 85-90°F (29-32°C). To create the gradient, use heat sources like ceramic heat emitters or heat mats.

lights: Since milk snakes mostly get their vitamin D from food, they don't need UVB lights. On the other hand, maintaining a consistent photoperiod and day-night cycle will help your snake's circadian rhythm and encourage its natural habits.

Select a substrate that will enable your milk snake to burrow and explore while also resembling its natural habitat. Aspen shavings, cypress mulch, coconut husk, or paper-based bedding are examples of appropriate substrate choices.

Hiding locations: To help your snake feel safe and comfortable, make sure its enclosure has a minimum of two hiding locations, one on the warm side and one on the chilly side. Natural objects, upturned flower pots, and commercial hides can all be used to create hiding places.

Ventilation: To preserve air quality and avoid the accumulation of moisture and smells, make sure the enclosure has enough ventilation. Ventilation panels or screened tops can aid in improving airflow and avoiding respiratory problems.

- Preventive Medicine and Veterinary Examinations

Preventive care and routine veterinarian exams are crucial for keeping your milk snake healthy and identifying any potential problems early. To guarantee your pet receives the best medical care possible, abide by these rules:

Annual Wellness Exams: Arrange for yearly wellness exams with a veterinarian specializing in reptiles to evaluate your snake's general health and to discuss any worries or inquiries you may have. Your veterinarian will do a physical examination, look for indications of disease or injury, and offer advice on preventive care during the examination.

Screening for Parasites: To check for internal parasites like worms or protozoa, consider doing yearly fecal parasite tests. To find and treat any parasitic illnesses, your veterinarian can take a fecal sample and run diagnostic testing.

Vaccinations: Although snakes are not usually vaccinated, talk to your veterinarian about any preventive care or treatments that may be suggested depending on the specific health requirements and risk factors of your snake.

Emergency Care: Become familiar with the typical symptoms of disease or distress in snakes, including hunger loss, aberrant behavior, lethargy, and respiratory problems. For advice and help in case of an emergency, get in touch with your veterinarian right away.

Maintaining records: Keep thorough records of any medical information pertaining to your snake, such as feeding regimens, shedding cycles, vet visits, and any alterations in look or behavior. You can use this information to keep an eye on the health of your snake and to communicate with your veterinarian.

- Symptoms of Disease and Veterinary Attention

Even with the best care you can provide them, milk snakes can occasionally become ill or have health problems. It's critical to keep an eye out for symptoms of disease and to take prompt action to get veterinarian care when necessary. Typical symptoms of disease in milk snakes include:

Appetite Changes: Abruptly losing your appetite or refusing to eat may be a sign of serious health problems such lung infections, metabolic disorders, or gastrointestinal parasites.

Abnormal Behavior: Your snake may be experiencing pain, stress, or sickness if it exhibits behavioral changes including lethargy, excessive hiding, or unusual aggression.

Breathing difficulties: Excessive mucus or saliva, open mouth breathing, wheezing, or difficulty breathing can all be signs of respiratory infections or other respiratory problems that need to be treated by a veterinarian.

Skin Abnormalities: Skin abnormalities such blistering, lesions, discolouration, or retained shed may be signs of underlying medical disorders like infections, wounds, or skin diseases.

Neurological Symptoms: Tremors, seizures, or unusual posture or movement are examples of symptoms that could point to neurological illnesses or other major health problems that need to be attended to by a veterinarian right away.

See your veterinarian right away for advice and support if you observe any symptoms of disease or unusual behavior in your milk snake. Your snake's chances of recovery can be increased by preventing the illness's progression through early detection and care.

To keep your milk snake healthy and happy, you must be vigilant, diligent, and provide proactive care. You can make sure your snake has a long, healthy, and happy life in captivity by giving it a balanced meal, enough water, ideal environmental conditions, and frequent veterinary checkups. Keep an eye on your snake's look and behavior, and if you spot any indications of disease or discomfort, get medical help right once. You can spend many years with your milk snake as a companion if you give it the right care and attention.

Chapter 6

Typical Problems and Their Fixes: Resolving Pet Care Concerns

As with any pet, taking care of a milk snake can be gratifying, but it also has its share of difficulties. Pet owners may run into a variety of challenges along the road, from health issues and behavioral issues to food issues and habitat issues. In this extensive guide, we will examine typical problems encountered by owners of milk snakes and offer workable methods to help resolve these concerns, guaranteeing the health and wellbeing of your companion.

- Feeding Difficulties

One of the most prevalent problems that owners of milk snakes encounter is feeding. Whether your snake exhibits finicky eating habits, refuses to eat, or regurgitates food, dealing with feeding issues takes time and cautious observation. The following list of typical feeding issues and possible fixes is provided:

Refusal to Eat: There are a few possible reasons why your milk snake might refuse to eat, including stress, the environment, or underlying medical conditions. Make sure your snake is comfortable in its enclosure by checking that the temperature, humidity, and hiding places are all at their ideal levels. To get your snake to eat, you should also experiment with providing a range of prey items and feeding it at different times of the day. Regurgitation: When a snake brings up its food soon after it has eaten, it is said to be regurgitating. This can happen for a variety of reasons, including stress, handling right after feeding, or the wrong size prey. Make sure the prey items are the right size for your snake's age and size, and refrain from handling it for at least 48 hours after feeding to prevent regurgitation.

See a reptile veterinarian if regurgitation continues in order to rule out any underlying medical conditions.

Picky Eating: Some milk snakes may exhibit picky eating behaviors as a result of developing preferences for particular prey items or feeding techniques. Try a variety of prey types, sizes, and feeding methods to see what suits your snake the best. To increase your snake's appetite, you can also try scenting prey items with natural scents, such as lizard or bird feathers.

- Environmental and Habitat Issues

Keeping your milk snake's habitat in good condition is crucial to its overall health and wellbeing. The comfort and behavior of your snake, however, can be affected by a number of external circumstances. The following list of typical habitat and environmental problems, along with their fixes:

Temperature Variations: Variations in the enclosure's temperature might cause stress to your snake and have an impact on its digestion and metabolism. Make sure the enclosure has a warm and a cool side, as well as a thermal gradient. To keep the temperature steady, employ heat sources and thermostats.

Humidity Imbalance: While too much humidity might encourage bacterial growth and respiratory infections, too little humidity can cause dehydration and shedding problems. To maintain ideal humidity levels, routinely check the levels and make necessary adjustments with water bowls, humid hides, or misting.

Inadequate Hiding Places: Not having enough hiding places might make milk snakes feel stressed and anxious, which makes them less active and reluctant to eat. Create a minimum of two hiding places using natural objects, overturned flower pots, or manufactured hides, one on the warm side and one on the cool side.

Substrate Problems: Inappropriate substrate might cause milk snake breathing problems, impaction, or skin irritation. Select a substrate, like aspen shavings, cypress mulch, coconut husk, or paper-based bedding, that allows your snake to burrow and explore in its natural habitat.

- Health Issues and Diseases

Like any reptile, milk snakes can suffer from a variety of ailments and health issues that could negatively affect their wellbeing. In order to properly handle health difficulties, it is imperative that individuals recognize the signs of illness and seek quick veterinarian care. The following are some typical health issues and how to fix them:

Snakes frequently develop respiratory infections, which can be brought on by poor husbandry, temperature swings, or pathogen exposure. Respiratory infections might manifest as wheezing, breathing difficulties, breathing through your mouth, or an abundance of mucus or saliva. For diagnosis and treatment, speak with a reptile veterinarian if you think your snake may have a respiratory infection.

Parasitic Infections: Your milk snake's health and vigor may be impacted by internal or external parasites like worms, mites, or ticks. Frequent veterinary examinations and fecal parasite screens can aid in the early detection and treatment of parasitic illnesses. For advice on the proper deworming or anti-parasitic medicines, speak with your veterinarian.

Skin Conditions: Incorrect shedding, humidity imbalances, or traumas can cause milk snakes to develop skin conditions like blistering, lesions, or retained shed. To encourage shedding, give your snake the right amount of humidity and soak times, and keep a close eye out for any irregularities on its skin. See a veterinarian that specializes in reptile care if you see any problems so they can assess and treat you.

Metabolic Disorders: The health and lifespan of milk snakes can be impacted by metabolic disorders such as obesity, vitamin deficits, or calcium inadequacy. To avoid metabolic imbalances, make sure your snake eats a balanced diet, stays well hydrated, and gets enough exposure to UVB or natural sunshine. For nutritional advice and, if necessary, supplementation, speak with a veterinarian that specializes in reptiles.

- Behavioral Issues

Inadequate environmental enrichment, mishandling, or stress can all lead to behavioral issues in milk snakes.

Finding the root cause and putting the right treatments in place are necessary for resolving these problems. The following list of typical behavioral issues along with their fixes:

Aggression: Stress, territoriality, or fear can all lead to aggressive behavior in milk snakes. When your snake is in a protective attitude or displaying aggressive tendencies, including hissing or striking, do not handle it. To ease tension and foster a sense of security, offer hiding places and enriching surroundings.

Hiding or Inactivity: Prolonged hiding or inactivity may be a sign of stress, disease, or unsuitable surroundings. In order to promote natural behaviors and activity, make sure the cage has enough hiding places, appropriate temperature gradients, and environmental enrichment. Keep an eye on your snake's behavior and contact a veterinarian specializing in reptiles if you have any concerns.

Escape Attempts: If their enclosure does not offer sufficient protection or if they are under stress, milk snakes, who are skilled escape artists, may try to break free. To stop escapes, make sure the cage contains barriers that cannot be broken through and safe locking systems. In order to decrease escape attempts, address any underlying sources of stress or discomfort.

Feeding Reluctance: Stress, environmental conditions, or medical conditions can all contribute to milk snake feeding reluctance. Make sure your snake's enclosure is comfortable, that there are enough hiding places, and that there are a variety of prey items available to pique its interest. Seek advice and examination from a reptile veterinarian if feeding resistance continues.

A milk snake needs commitment, perseverance, and proactive problem-solving to handle typical difficulties and guarantee your pet's health and wellbeing. Through comprehension of the fundamental reasons behind feeding challenges, habitat malfunctions, health issues, and behavioral issues, you may put suitable remedies into place and give your snake the finest care possible. Your milk snake's health and pleasure can only be sustained for many years with consistent observation, monitoring, and veterinarian care. You can have a happy and satisfying relationship with your milk snake as a treasured pet if you give it the right care and attention.

Chapter 7

Fun Exercises and Enhancement: Maintaining the Stimulation of Your Milk Snake

Even while milk snakes might not be as social as some other pets, it's still important to give them stimulation and enrichment to support their physical and mental well. Engaging in enrichment activities can assist avoid boredom, promote innate habits, and offer chances for physical activity and discovery. We'll look at a variety of entertaining activities and enrichment suggestions in this extensive guide to keep your milk snake interested and active.

- The Value of Enhancement

The wellbeing and well-being of captive snakes, especially milk snakes, depend on enrichment. Snakes exhibit a range of activities in their natural habitat, such as hunting, exploring, basking, and concealing. Snakes kept in captivity would not have as many opportunities to exhibit these natural behaviors, though. By simulating elements of the snake's natural habitat and stimulating its senses, enrichment activities seek to both physically and mentally stimulate the animal.

Your milk snake's daily routine can be made less stressful, less boring, and more conducive to overall wellbeing by adding enrichment. Through opportunities for interaction and positive reinforcement, enrichment activities also contribute to the development of a deep link between you and your snake.

- Enhancement of Environment

The goal of environmental enrichment is to improve the physical surroundings of the snake in order to pique its curiosity and promote natural behaviors. To improve the environment of your milk snake's enclosure, consider the following ideas:

Milk snakes are semi-arboreal and like to climb structures. Give your snake some branches, logs, or climbing vines to explore and clamber on. To keep climbing structures from collapsing or hurting your snake, make sure they are fastened firmly.

Hiding Spots: To increase security and lower anxiety, provide a number of hiding places within the enclosure. Make hiding places on the warm and cool sides of the enclosure with store-bought hides, converted flower pots, or natural items like boulders or cork bark.

In order to arouse your snake's senses, add new smells, textures, and objects to its habitat. To keep things interesting, you may add leaves, branches, or other natural materials to the cage. You can also switch up the accessories and decorations from time to time.

Substrate Variety: Try a range of substrate kinds to stimulate the senses and promote natural burrowing habits. Milk snakes can be housed on aspen shavings, cypress mulch, coconut husk, or bedding made of paper.

Environmental Changes: To add variation and surprise, periodically modify the enclosure's layout or add new objects and decorations. These modifications may pique your snake's interest and motivate exploration.

- Providing Enrichment

In order to arouse your snake's hunting instincts and encourage cerebral engagement, you should feed it in a way that offers both diversity and challenge. Here are some suggestions for enriching your milk snake's diet:

Feeding Puzzles: To make mealtimes more interesting and challenging for your snake, use feeding puzzles or enrichment tools. You can conceal prey items for your snake to find behind rocks or inside cardboard tubes, or you can put prey items inside a PVC pipe that has holes punched through it.

Scent Enrichment: You can pique your snake's interest in hunting and promote eating by scent-enriching prey items with natural scents, such lizard or bird feathers. Before giving prey items to your snake, you can dust them with bird feathers or rub them against a piece of lizard skin.

Live Prey Simulation: Feeding tongs or fishing line can be used to replicate the movement of live prey. To foster hunting behavior and arouse your snake's senses, gently jiggle the prey item in front of it.

Food Dispensing Toys: Give your reptiles puzzle feeders or food dispensing toys to keep their minds active while they eat. These toys encourage problem-solving and organic foraging behaviors by requiring your snake to manipulate and interact with the device in order to access the food.

Feeding Time Variability: To avoid monotony and maintain your snake's interest, change up the time, place, and manner in which meals are presented. To test your snake's hunting prowess, provide prey items at different times of the day or night, conceal prey items inside the enclosure, or present prey items in unique ways.

- Enhancement of Behavior

The goals of behavioral enrichment are to encourage innate tendencies and to offer chances for mental and physical exercise. Here are some suggestions for your milk snake's behavioral enrichment:

Explore: Give your snake the freedom to go about in a secure, monitored setting when it's not in its enclosure. Reptile-safe barriers can be used to create a snake-proofed area, or you can keep an eye on your snake in a secure room while it roams free.

Offer chances for sensory stimulation by acquainting people with new sights, sounds, and sensations. You can play soothing music or natural noises, put the enclosure next to a window so your snake can see outdoor activity, or give your snake textured surfaces to explore.

Basking and Sunlight Exposure: Position the enclosure close to a window or give it access to a UVB light source to create opportunities for basking and natural sunlight exposure. UVB rays and natural sunshine are vital to the general health and wellbeing of your snake.

Even though milk snakes are solitary creatures, on occasion handling and engagement with their owner can foster social enrichment and the development of confidence and trust. Allow your snake to explore and engage with you at its own speed while handling it with gentleness and respect.

Training and Targeting: Use positive reinforcement strategies to teach your snake to obey basic orders or targets. You can train your snake to perform simple tasks or behaviors with the assistance of a target stick or feeding tongs, rewarding it with praise or food.

- Play and Conversation

Even though milk snakes don't play like dogs or cats do, they can still gain by socializing and engaging in interactive play with their owners. Observe these methods to engage with your milk snake and strengthen your bond:

Gentle handling sessions on a regular basis offer your snake the chance to socialize and strengthen their bond. In a secure and supervised setting, let your snake explore and engage with you while handling it with composure and assurance.

Exploration Time: Provide your snake with safe and secure surroundings to explore outside of its habitat while being supervised. Reptile-safe barriers can be used to create a snake-proofed area, or you can keep an eye on your snake in a secure room while it roams free.

Training Exercises: Use positive reinforcement methods to teach your snake basic habits or tricks. You may teach your snake to explore, climb, or engage with items in its surroundings by using basic commands or target training.

Interactive Toys: To stimulate the mind and promote exercise, give interactive toys or enrichment items made specifically for reptiles. These toys can be puzzle feeders, climbing frames, or items that promote discovery and communication.

Bonding Time: Sit close to your snake's enclosure, converse with it, or give it soft pats and strokes to spend quality time together. Even though they might not be as affectionate as mammals, snakes can nonetheless gain from and enjoy good interactions with their owners.

- Safety Points to Remember

Prioritize your own and your pet's safety when offering enrichment and participating in interactive activities with your milk snake. Observe these safety precautions to guarantee a satisfying and joyful experience:

Always keep an eye on your snake when it is being explored or interacted with outside of its enclosure to avoid mishaps or escapes. Pay careful attention to your snake's behavior and offer assistance and direction as required.

Safe Enclosure: To avoid mishaps or escapes, make sure the snake's enclosure is safe and impenetrable. Make sure that doors and lids are locked, and periodically check the enclosure for wear or damage that might affect its structural integrity.

Gentle Handling: Treat your snake with deference and gentleness; steer clear of abrupt movements or rough treatment that might aggravate or injure it. Give your snake the support it needs and let it come to you and explore at its own speed.

Regularly inspect your snake's health to look for any indications of illness or damage. If you notice any changes in your reptile's appearance, behavior, or appetite, speak with a veterinarian about your concerns.

Positive Reinforcement: To promote desired behaviors and interactions with your snake, use positive reinforcement strategies like praise or food rewards. Reward or punishment should be avoided since they can aggravate your pet and undermine your mutual trust.

The promotion of milk snakes' health, happiness, and general well-being in captivity depends on enrichment and stimulation. You may give your snake a stimulating and rewarding environment that supports its natural activities and promotes positive contact with you as its owner by including interactive playtime, feeding enrichment, behavioral enrichment, and environmental enrichment into its daily routine. As you appreciate your milk snake as a pet, never forget to put safety first, honor your snake's own preferences and boundaries, and relish the special and fulfilling relationship you have with it. You can give your snake a long, happy life filled with enrichment if you give it the right care.

Chapter 8

Common Questions and Answers (FAQs) about Milk Snakes

Popular pet reptiles, milk snakes are prized for their vivid colors, placid disposition, and low maintenance needs. Like any pet, though, potential owners may have concerns regarding their upkeep, mannerisms, and suitability as companions. We will answer some of the most common queries regarding milk snakes in this extensive guide, offering precise details and helpful guidance for both present and potential owners.

1. Describe a milk snake.

Native to both North and South America, milk snakes (Lampropeltis triangulum) are non-venomous constrictor snakes. They are characterized by their remarkable coloring, which usually consists of vivid bands of red, black, yellow, or white. They are members of the colubrid family.

2. Are pet milk snakes healthy?

Yes, reptile lovers of all skill levels can have wonderful pets in milk snakes. Generally speaking, they require less care than other reptiles and are calm and easy to handle. To flourish in captivity, they do, however, need the same kind of care, attention, and environment as any other pet.

3. How big can milk snakes get?

Adult milk snakes can grow to an average length of 2 to 4 feet (60 to 120 centimeters), making them generally medium-sized snakes. Individual specimens and some subspecies, however, may grow larger or smaller according on environmental circumstances, food, and genetics.

4. Milk snakes consume what?

Milk snakes are carnivorous reptiles that mostly consume rodents, such as mice and rats. While adult snakes can eat larger prey items like adult mice or small rats, hatchlings and juveniles usually eat smaller prey items like pinky mice or fuzzies. It's critical to give prey items that are proportionate to your snake's size and age.

5. How frequently should milk snakes be given food?

Milk snake feeding frequency is influenced by age, size, and metabolism, among other things. Juveniles and hatchlings may need to be fed more frequently, usually every five to seven days, in order to sustain their rapid development and growth. Feeding should only take place every 7 to 10 days for adult snakes, or more often if necessary to keep them at a healthy weight.

6. Are UVB lights necessary for milk snakes?

Since milk snakes mostly get their vitamin D from food rather than UVB radiation, they do not require UVB lights. On the other hand, maintaining a regular photoperiod and day-night cycle will help your snake's circadian rhythm and encourage its natural habits.

7. Do cowslips require a heating lamp?

Even though milk snakes can regulate their body temperature, an enclosure must include a heat source in order to create a thermal gradient. On one side of the cage, a warm basking area between 85 and 90°F (29 and 32°C) can be created using a heat lamp or heat mat. It is recommended to keep the enclosure's cold side at a slightly lower temperature, between 75 and 80°F (24 and 27°C).

8. How frequently do milk snakes molt?

When they mature, milk snakes normally shed their skin every 4 to 8 weeks, depending on their age, growth pace, and surroundings. Snakes shed their skin naturally, which is an important process that helps them to get rid of old, dead skin cells and show new, vivid skin underneath.

9. What symptoms indicate that my milk snake is about to shed?

Here are some telltale signs that your milk snake is about to shed:

1. hazy or opaque eyes; also referred to as "blue" eyes
2. dull or fading color of the skin
3. Reduced desire to eat
4. heightened energy or agitation
5. rubbing up against the enclosure's surfaces or objects

10. Is it possible for milk snakes to coexist in the same enclosure?

Although milk snakes are solitary creatures in general, some circumstances can allow them to live in harmony within the same habitat. Cohabitation, however, raises the possibility of stress, aggressiveness, and competition, particularly during the feeding or breeding seasons. In order to avoid disputes when housing many milk snakes together, it's critical to offer enough room, hiding places, and supervision.

11. Are milk snakes noisy animals?

No, vocalizations or noises are not typically made by milk snakes. They are usually solitary, silent creatures that mostly use body language to communicate, like hissing or striking when they feel frightened or anxious.

12. What is the lifespan of milk snakes?

Milk snakes can survive in captivity for up to 20 years or more with the right maintenance and care. You may extend the life and improve the quality of your snake's habitat, diet, and activities for enrichment. You should also provide regular veterinary treatment and a balanced diet.

13. Do milk snakes have a violent nature?

Although milk snakes are generally calm and non-aggressive animals, each snake's temperament can differ based on a variety of factors, including handling, genetics, and environmental circumstances. Even though they might hiss or strike in response to stress or threat, they are usually tolerant of handling and, with the right socialization, can grow accustomed to human interaction.

14. How do I deal with a milk snake?

It's important to approach a milk snake with confidence and calmness, supporting its body appropriately to avoid stress and injury. Just behind the head, use one hand to support the front third of the snake's body; use the other to support the middle and tail. It is best to let the snake move freely and investigate your hands and arms at its own speed, avoiding abrupt movements or loud noises that could frighten it.

15. Can a person train a milk snake?

Although they might not be as teachable as certain other pets like dogs or cats, milk snakes can recognize their owners and react well to tactics of positive reinforcement. Simple commands or target training are examples of training exercises that might help your snake become more engaged and intelligent.

A fascinating pet reptile, milk snakes are prized for their beauty, composure, and ease of maintenance. You may provide your milk snake an appropriate environment and enjoyable experiences by being aware of their nutritional preferences, temperament, and habitat requirements. Building a happy and meaningful relationship with your milk snake may be a rewarding and pleasurable experience, regardless of your level of expertise as a snake keeper. You can have a treasured pet and many years of friendship with your milk snake if you give it the right care, attention, and dedication.

Chapter 9

In conclusion, relishing the special pleasures of owning a milk snake

Any age group of reptile enthusiasts can find great satisfaction and enrichment in being the owner of a milk snake. Milk snakes have won over the hearts of innumerable pet lovers worldwide with their brilliant colors, placid disposition, intriguing activities, and comparatively low maintenance care needs. As we draw to a close this thorough guide on milk snake ownership, let's consider the special pleasures and advantages of taking care of these fascinating reptiles.

1. Grace and Variety

The vivid and colorful appearance of milk snakes is one of its most remarkable characteristics. A sight to behold, milk snakes have strong stripes of red, black, and yellow or white. The distinct hue and pattern of every subspecies and individual specimen contribute to the attraction and diversity of these stunning reptiles. There is a milk snake type to fit every aesthetic style, whether you like the subtle grace of the Sinaloan milk snake or the stark contrast of the Honduran milk snake.

2. Calm and Capable Ecology

Because of their well-known placid and manageable disposition, milk snakes are perfect companions for reptile aficionados of all skill levels. In contrast to several other snake species, milk snakes can get accustomed to human contact with adequate socialization and are often tolerant of handling. The serene and kind nature of milk snakes makes them a pleasure to handle and engage with, regardless of your level of experience as a reptile enthusiast.

3. Possibilities for Education

Having a milk snake offers parents and kids alike extra educational possibilities. We have the chance to learn about the natural history, behavior, and ecology of these amazing reptiles as their custodians. Whether studying their natural environments or monitoring their hunting and eating habits in captivity, keeping a milk snake in your possession can provide an engaging educational opportunity that can help you develop a greater respect and comprehension of the natural world.

4. Therapeutic Advantages

Humans have been demonstrated to benefit therapeutically from interactions with animals in a variety of ways, including decreased stress, anxiety, and sadness. Having a milk snake as a pet can offer chances for mindfulness and relaxation, as well as a sense of purpose and friendship. A milk snake's soothing presence and the tactile experience of handling and tending to a live being can have a profoundly positive impact on one's mental and emotional health.

5. Connectivity and Friendship

Even though they might not show as much social bonding as mammals, milk snakes are nonetheless able to develop deep bonds with their owners. Regular handling, constructive criticism, and engaging playing help owners establish rapport and trust with their milk snakes, which promotes respect and companionship. An owner-snake bond may be quite fulfilling, whether it is through peaceful times spent together, feeding the snake prey, or watching the snake explore its habitat.

6. Awareness of Conservation

Having milk snake can also help spread awareness of the value of protecting reptile habitat and conservation. Through gaining knowledge about the challenges that wild snake populations face, such as habitat loss, pollution, and poaching, owners can develop into champions for environmental responsibility. The conservation of milk snakes and their natural habitats can be aided by funding conservation projects, teaching people about the importance of snakes in ecosystems, and engaging in morally responsible and sustainable pet ownership.

7. Feeling of Obedience

Taking care of a milk snake requires owners to be dedicated and responsible because they are taking on the task of providing for the requirements of another living being. Owning a milk snake demands commitment and meticulousness, from upholding an appropriate environment and feeding it a balanced food to keeping an eye out for any symptoms of illness and getting veterinary care when necessary. But the benefits of having a pet responsibly are enormous; one such benefit is seeing your snake grow and prosper.

8. Feeling of Wonder

Owning a milk snake, above all, encourages awe and admiration for the intricacy and beauty of the natural world. Owners of milk snakes are filled with wonder and interest, whether it's from studying about their evolutionary adaptations, watching their hunting and eating habits, or marveling at their beautiful pigmentation. Every day, the chance to tend to and engage with these fascinating animals serves as a reminder of the glories of the natural world and the interdependence of all living things.

In summary

To sum up, having a milk snake as a pet is a singular and rewarding experience that combines companionship, beauty, and learning possibilities. The remarkable looks, kind nature, and intriguing habits of milk snakes can entice you to keep them as pets or companions. Owners may guarantee that their milk snake has a happy, healthy, and meaningful life in captivity by providing appropriate care, enrichment activities, and a suitable habitat.

Remember to treat your newfound experience as a milk snake owner with compassion, curiosity, and respect for these amazing reptiles. Spend some time learning about their natural history, temperament, and care needs. When in doubt, seek advice from vets and seasoned reptile keepers. You can experience the special pleasures of owning a milk snake for many years to come if you have patience, commitment, and a sense of wonder.

www.ingramcontent.com/pod-product-compliance
Lightning Source LLC
Chambersburg PA
CBHW050326230526
45471CB00005B/2374